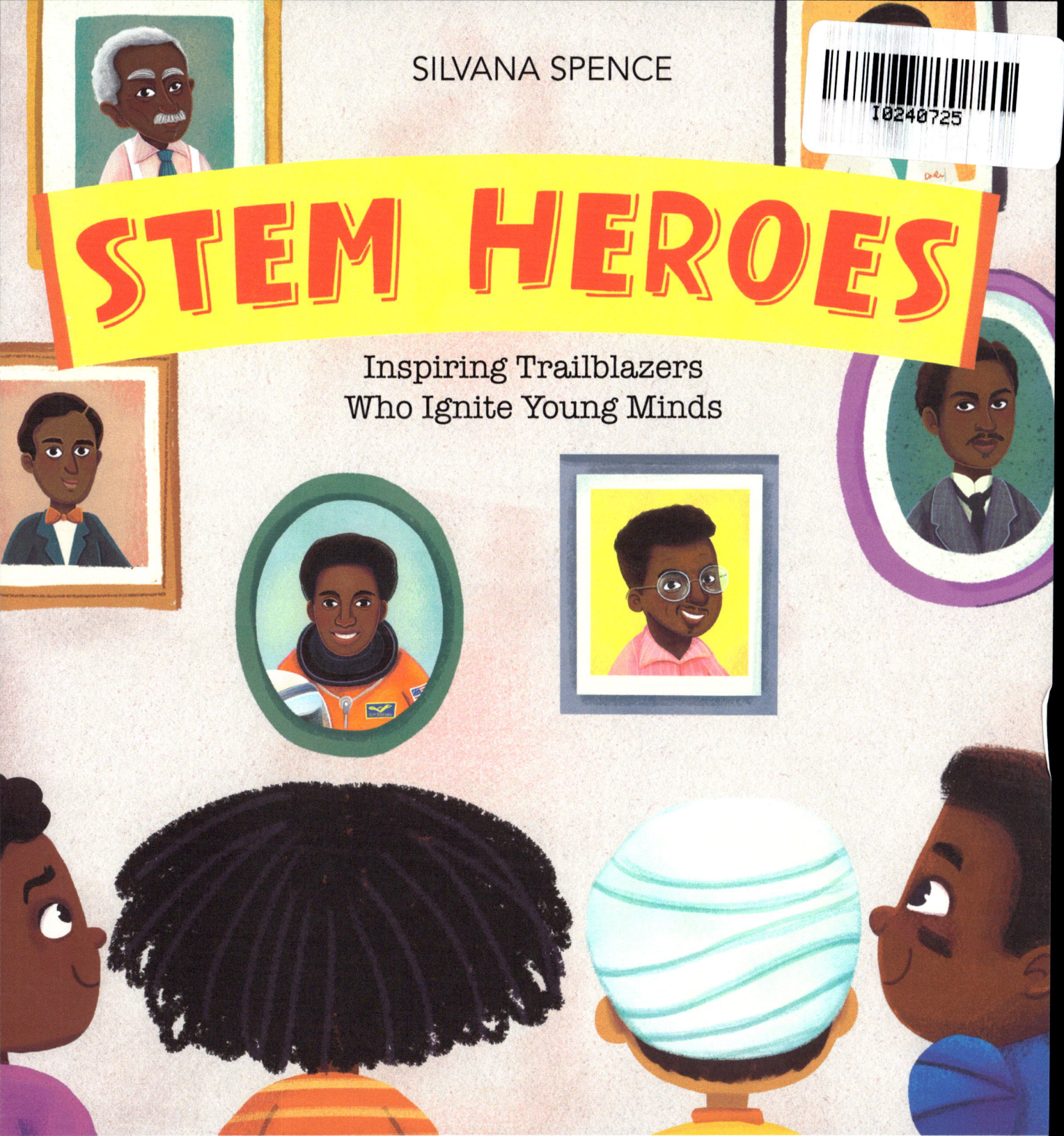

STEM Heroes: Inspiring Trailblazers Who Ignite Young Minds.
Copyright ©2024 by Silvana Spence.
All rights reserved.

No part of this book may be reproduced in any written, electronic, recording, or photocopying form without written permission of the author, Silvana Spence and Victoria Spence, or the publisher, The STEAM SQUAD, LLC.
Books may be purchased in quantity and /or special sales through the website www.bellathescientist.com.

Published by: The STEAM SQUAD, LLC.
Illustrated by: Claudia Marianno
First edition, 2024.

Dedication

There is a world of possibilities for what you can become one day.
Read, research, and find what you can do best.
Go for it! I believe in you!

Special Thank you to

De'Shaan Dixon

Anthony B. Miles

RAM - Riverside Arts Market

Jax Melanin Market

We appreciate your support to children's literacy and STEM.

There are many options for STEM careers.
I choose the one inspired by pioneers.
With persistence, I strive for an important degree,
The best STEM professional I'm confident I'll be.

I can be an astronaut like Dr. Guion Stewart Bluford Jr., the first African-American in space. He rode on the Space Shuttle Challenger as part of the STS-8 mission, which launched on August 30, 1983. The crew deployed an Indian communications and weather satellite. The Challenger returned to Earth on September 5, 1983.

I can be a pilot like <u>Colonel Charles McGee.</u> He served with distinction in the United States Air Force as part of the Tuskegee Airmen. A national hero, he flew over 100 combat missions during World War II.

I can be a chemist like <u>George Washington Carver</u>. An agricultural scientist, he invented hundreds of products made from peanuts, sweet potatoes, and soybeans, which are now used worldwide. His inventions included the Jesup Wagon, a mobile school that brought education to rural areas.

I see what I can be.
These **STEM** heroes lead the way for me.
No matter the obstacles I face,
I'll succeed in life with grace.

I can be a computer scientist like Dr. Mark Dean, an African-American inventor who, as an IBM Fellow, paved the way for the development of the ISA bus connector. He also led the design team that built the first gigahertz computer processing chip. Dr. Dean holds three of IBM's original nine PC patents and pioneered personal computing.

I can be a doctor like Benjamin Solomon Carson Sr., A renowned neurosurgeon who made history by successfully separating conjoined twins in 1987. He served as the Secretary of Housing and Urban Development of the United States from 2017 to 2021.

The engineering field is a vast domain,
and I'll strive for excellence and expertise to gain.
Nuclear, electrical, or civil to pursue,
I'll make the best choice and carry it through.

I can be a nuclear engineer like Lonnie George Johnson, who developed a nuclear-powered probe for NASA. He also invented the Super Soaker water gun, one of the world's bestselling toys.

PATENTS

PORTABLE MULTIMEDIA PROJECTION

DIGITAL DISTANCE MEASURING INSTRUMENT

HAIR CURLER APPARATUS

I can be a video game designer like Jerry Lawson. An engineer and inventor, he designed a new video game console that allowed players to switch out cartridges. Later, he founded Video Soft, Inc., the first video game company in the United States, by an African-American.

I can be an inventor like <u>Garrett Augustus Morgan Sr.</u> He was an African-American entrepreneur and community leader who invented the traffic signal and the safety hood, along with an early version of the gas mask and a hair-straightening solution.

Visions unfold my ancestors' legacy in me,
They paved the way from struggle to victory.
Challenges may come in this success race,
With determination, I'll claim my place.

I can be a hematologist like <u>Dr. Charles Richard Drew</u>, known for the blood banks he started early in World War II. He was the first director of the American Red Cross after his accomplishment of developing the first mobile blood-collecting unit in 1941. The unit allowed blood plasma to be delivered to the war front for blood transfusions that saved many lives during World War II and even now.

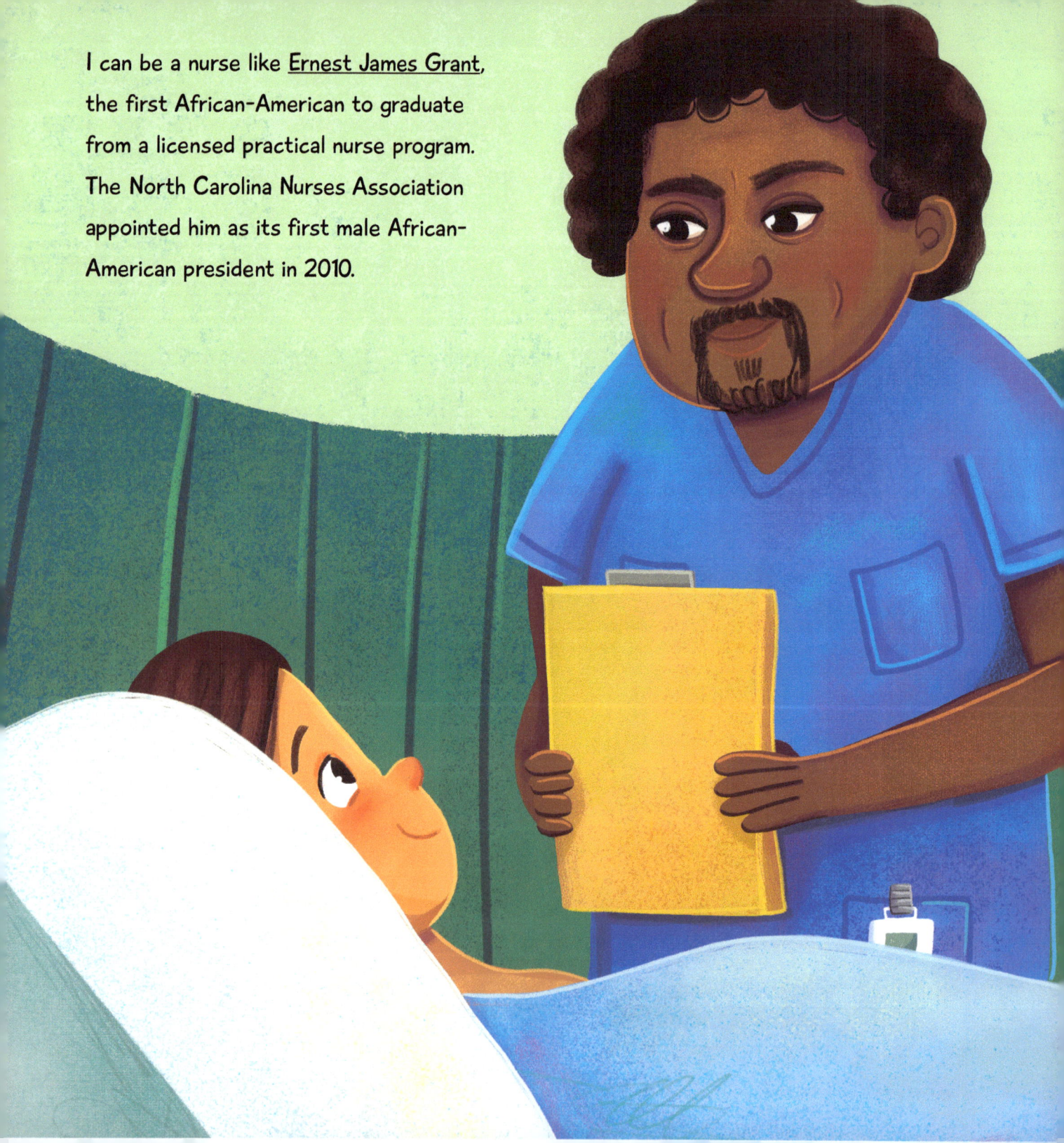

I can be a nurse like <u>Ernest James Grant</u>, the first African-American to graduate from a licensed practical nurse program. The North Carolina Nurses Association appointed him as its first male African-American president in 2010.

I can be an ophthalmologist like Dr. David Kearny McDonogh, a specialist in ophthalmology, the study of the eyes, and otolaryngology, the study of the ears, nose, and throat. He graduated from Lafayette College in 1844, the first African-American to do so, and was in the top three in his class.

Your greatness inspires me to walk on the right road.
Your resilience requires me to be strong on this lode.
The trailblazers in STEM I respect now, and then.

I can be a marine biologist like <u>Professor Robert Kent Trench</u>, one of the world's leading experts on symbiosis in coral reefs. He also served as a founding board member of the Global Coral Reef Alliance.

I can be a veterinarian like Augustus Nathaniel Lushington, the first Black veterinarian in the United States. Dr. Lushington moved from Trinidad to the United States when he was 20. He enrolled in Cornell University, where he graduated with a degree in agriculture in 1894. He earned his doctorate from the University of Pennsylvania in 1897.

I can be a biologist like <u>Ernest Everett Just</u>. An African-American biologist and educator. He graduated from Dartmouth College with honors in Zoology in 1907 and a post-graduate in Biology at the University of Chicago. He focused his research on marine invertebrate eggs and fertilization.

To mirror your greatness, is what I aspire.
Your perseverance is what I admire.
Adversities you faced to get to victory,
A STEM- Hero means you made history

References

(n.d.). Lonnie Johnson is the inventor of the Super Soaker. Retrieved January 5, 2024, from

    http://lonniejohnson.com/

(n.d.). Dr. David Kearney McDonogh, M.D. Retrieved January 5, 2024, from

    http://www.davidkearneymcdonoghmd.com/

Archive, H. (n.d.). Black Scientists and Inventors | Black History Month. National Geographic Kids. Retrieved January 5, 2024,

    from *https://kids.nationalgeographic.com/science/article/black-inventors-and-pioneers-of-science*

Benjamin, W. (2007, January 18). Robert Tanner Freeman (1846-1873) •. Blackpast. Retrieved January 5, 2024, from

    https://www.blackpast.org/african-american-history/freeman-robert-tanner-1846-1873/

Black History Month 2021 | Coastal and Marine Laboratory. (2021, February 10). FSU Coastal and Marine Laboratory. Retrieved January 5, 2024, *from https://marinelab.fsu.edu/news-at-fsucml/black-history-month-2021/*

Black History Month 2021 Health Pioneers: An Interview with Dr. Ernest J. Grant | NC AAHC. (n.d.). North Carolina African American Heritage Commission. Retrieved January 5, 2024, from https://aahc.nc.gov/resources/black-history-month-2021/black-history-month-2021-health-pioneers-interview-dr-ernest-j-grant

Charles Richard Drew. (n.d.). American Chemical Society. Retrieved January 5, 2024,

    from *https://www.acs.org/education/whatischemistry/african-americans-in-sciences/charles-richard-drew.html*

Dr. Charles Drew. (n.d.). Profiles in Science. Retrieved January 5, 2024, from

    https://profiles.nlm.nih.gov/spotlight/bg/feature/biographical-overview

Dr. Mark Dean: Computer Inventions — Famous Black Inventors. (n.d.). Famous Black Inventors. Retrieved January 5, 2024,

    from *https://www.black-inventor.com/dr-mark-dean*

Garrett Morgan is the inventor of the Traffic Light & Gas Mask. (n.d.). Biography (Bio.). Retrieved January 5, 2024,

    from *https://www.biography.com/inventors/garrett-morgan*

Garrett Morgan - Kids. (n.d.). Britannica Kids. Retrieved January 5, 2024,

    from *https://kids.britannica.com/kids/article/Garrett-Morgan/443679*

Goreau, T. (2021, April 28). Robert Kent Trench: In Memoriam. Global Coral Reef Alliance. Retrieved January 5, 2024, from

    https://www.globalcoral.org/robert-kent-trench-in-memoriam/

Jerry Lawson - Video Games, Death & Life. (2014, April 3). Biography (Bio.). Retrieved January 5, 2024,

    from https://www.biography.com/inventor/jerry-lawson

Koplin, R. S. (2016, October 6). America's First African-American Eye Specialist: David K. McDonogh, MD. American Academy of Ophthalmology. Retrieved January 5, 2024, from https://www.aao.org/senior-ophthalmologists/scope/article/david- mcdonogh-african-american-eye-specialist

Koplin, R. S. (2016, October 6). America's First African-American Eye Specialist: David K. McDonogh, MD. American Academy of Ophthalmology. Retrieved January 5, 2024, from https://www.aao.org/senior-ophthalmologists/scope/article/david- mcdonogh-african-american-eye-specialist

Library Guides: Black Veterinary History: Augustus N. Lushington. (2023, November 30). Library Guides. Retrieved January 5, 2024, from https://libraryguides.missouri.edu/AfricanAmericanVeterinaryHistory/Lushington

Logsdon, J. M. (2023, December 25). Guion Bluford | Biography, Spaceflights, & Facts. Britannica. Retrieved January 5, 2024, from https://www.britannica.com/biography/Guion-Bluford

Lonnie Johnson | The National Inventors Hall of Fame. (2024, January 2). National Inventors Hall of Fame®. Retrieved January 5, 2024,

    from https://www.invent.org/inductees/lonnie-johnson

McSwine, D. (2023, November 7). .,. ., - YouTube. Retrieved January 5, 2024, from

    https://link.springer.com/article/10.1007/s13199-021-00817-w

Pollitt, P. (n.d.). Ernest J. Grant | North Carolina Nursing History. North Carolina Nursing History. Retrieved January 5, 2024,

    from https://nursinghistory.appstate.edu/biographies/ernest-j-grant

Robert Tanner Freeman, DMD, Class of 1869. (n.d.). Perspectives Of Change. Retrieved January 5, 2024,

    from https://perspectivesofchange.hms.harvard.edu/node/43

Rogers, K. (n.d.). Ben Carson | Biography & Facts. Britannica. Retrieved January 5, 2024,

    from https://www.britannica.com/biography/Ben-Carson

Saucier, J. (2013, December 2). Collection Documents the Career of Video Game Pioneer Jerry Lawson - The Strong National Museum of Play. Strong Museum of Play. Retrieved January 5, 2024,

    from https://www.museumofplay.org/blog/collection- documents-the-career-of-video-game-pioneer-jerry-lawson/

About the Author

Originally from Brazil, **Silvana Spence** is a mom, author, and educator who loves science and literacy. She is the author of Bella the Scientist Goes to Outer Space, which she co-wrote with her daughter Isabella and features her daughter Victoria. She also wrote " Here We Go," an easy-reading book series in collaboration with the Library Foundation of Jacksonville. Silvana loves spending time writing, traveling, and enjoying her family. As a strong advocate for STEAM (Science, Technology, Engineering, Arts, and Math), She wants her readers to be represented and motivated to be the best they can be.

Website: *www.bellathescientist.com*

Social media: *www.instagram.com/mrs.s4success*
www.instagram.com/bellathescientist

www.ingramcontent.com/pod-product-compliance
Lightning Source LLC
Chambersburg PA
CBHW061402090426
42743CB00002B/110